Materials

Wood

Cassie Mayer

Heinemann Library
Chicago, Illinois

Customer Service 888-454-2279
Visit our website at www.heinemannraintree.com

Designed by Joanna Hinton-Malivoire
Printed in China by South China Printing Company Limited

12 11 10 09 08
10 9 8 7 6 5 4 3 2 1

ISBN-13: 978-1-4329-1617-6 (hc)
ISBN-13: 978-1-4329-1626-8 (pb)

The Library of Congress has cataloged the first edition as follows:
Mayer, Cassie.
   Wood / Cassie Mayer. -- 1st ed.
      p. cm. -- (Materials)
   Includes bibliographical references and index.
   ISBN 978-1-4329-1617-6 (hc) -- ISBN 978-1-4329-1626-8 (pb) 1. Wood--Juvenile literature. I. Title.
   TA419.M463 2008
   620.1'2--dc22
                              2008005577

Acknowledgments
The author and publisher are grateful to the following for permission to reproduce copyright material: ©Corbis pp. **5** (Kevin Fleming), **20** (Brand X); ©Heinemann Raintree pp. **6**, **9**, **10**, **11**, **12**, **13**, **15**, **16**, **17**, **21**, **22BL**, **22BR**, **23B**, **23T** (David Rigg); ©istockphoto pp. **18**, **22TR** (Michael Braun)' ©Shutterstock pp. **4** (Elena Elisseeva), **7** Olga Zaporozhskaya), **8** (Mikhail Olykainen), **19** (Semen Lixodeev), **22TR** (Olga Zaporozhskaya), **23M** (Elena Elisseeva).

Cover image used with permission of ©Getty Images (Chuck Kuhn Photography, Inc.). Back cover image used with permission of ©istockphoto (Michael Braun).

Every effort has been made to contact copyright holders of any material reproduced in this book. Any omissions will be rectified in subsequent printings if notice is given to the publisher.

Contents

What Is Wood?

Wood is in nature.

Wood is from trees.

Wood can be hard.

Wood can be smooth.

Wood can be heavy.

Wood can be light.

What Happens When Wood Is Heated?

Wood can be heated.

Wood can catch fire.

Wood can burn.

Wood can turn to ash.

How Does Water Change Wood?

Wood can be changed by water.

Wood can be changed over time.

Wood can become soft.

Wood can become rotten.

How Do We Use Wood?

Wood can be used to build.

Wood can be used to make paper.

Wood can be used to make fires.

Wood can be used to make
many things.

Things Made of Wood

▲houses

▲tables and chairs

▲sculptures

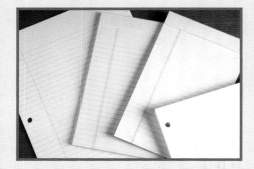

▲paper